ANIMALS WITHOUT BRAINS!

SEA CUCUMBERS ARE BRAINLESS!

By Amanda Vink

Gareth Stevens PUBLISHING

Please visit our website, www.garethstevens.com. For a free color catalog of all our high-quality books, call toll free 1-800-542-2595 or fax 1-877-542-2596.

Library of Congress Cataloging-in-Publication Data

Names: Vink, Amanda, author.
Title: Sea cucumbers are brainless! / Amanda Vink.
Description: New York : Gareth Stevens Publishing, [2020] | Series: Animals without brains! | Includes index.
Identifiers: LCCN 2019020402| ISBN 9781538245903 (paperback) | ISBN 9781538245927 (library bound) | ISBN 9781538245910 (6 pack)
Subjects: LCSH: Sea cucumbers–Juvenile literature.
Classification: LCC QL384.H7 V56 2020 | DDC 593.9/6–dc23
LC record available at https://lccn.loc.gov/2019020402

Published in 2020 by
Gareth Stevens Publishing
111 East 14th Street, Suite 349
New York, NY 10003

Designer: Sarah Liddell
Editor: Amanda Vink

Photo credits: Cover, p. 1 Jung Hsuan/Shutterstock.com; background used throughout Sanjatosic/Shutterstock.com; p. 5 Erin Wildermuth/Shutterstock.com; p. 7 orlandin/Shutterstock.com; p. 9 e2dan/Shutterstock.com; p. 11 Damsea/Shutterstock.com; p. 13 Jaya Kesavan/Shutterstock.com; p. 15 tank200bar/Shutterstock.com; p. 17 Laura Dinraths/Shutterstock.com; p. 19 Ethan Daniels/Shutterstock.com; p. 21 unterwegs/Shutterstock.com.

Printed in the United States of America

CPSIA compliance information: Batch #CW20GS: For further information contact Gareth Stevens, New York, New York at 1-800-542-2595.

CONTENTS

Boldface words appear in the glossary.

No Brain!

You may not be surprised to learn that a sea cucumber doesn't have a brain. After all, they look a little like actual cucumbers. Sea cucumbers don't have hearts or lungs either! How do they live without these body parts?

Invertebrates

Sea cucumbers don't have a **skeleton** or backbone. This makes them part of the invertebrate family, a group that's part of the **animal kingdom**. There are many invertebrates on Earth. In fact, 97 percent of all animals fall into this group!

A Sea Cucumber's Body

A sea cucumber's body is one long tube with small tube-like feet. They have no face and no eyes, and they have a mouth surrounded by between 8 and 30 feet. Their soft body is covered in rough skin.

Life at the Bottom

There are 1,250 known species, or kinds, of sea cucumbers. They spend their lives at the very bottom of the ocean, inching along slowly using their tube feet. They eat tiny bits of algae, very small sea creatures, and food waste.

Nature's Garbage Disposal

Sea cucumbers help to clean up the ocean! A sea cucumber's waste becomes the food for **bacteria**. It's recycled back into the **ecosystem**. Some sea cucumbers catch falling particles in the water, and others move through the sand.

A Simple Nervous System

Brains take a lot of energy to power—and some creatures, such as the sea cucumber, don't need to think so much. A sea cucumber's simple **nervous system** takes in what's happening around them. It helps the sea cucumber respond to movement and changes in light.

The Nerve!

Nerves are part of the nervous system. They are cells that collect and carry messages all over the body. A nerve ring surrounds a sea cucumber's mouth. Sea cucumbers have a nerve net under their skin, which gives them a sense of touch.

Defense

When a sea cucumber is in danger, there aren't many options. When a predator attacks, some species of sea cucumber save themselves by creating a distraction. They tighten up and shoot out body parts from inside their body. These parts grow back!

Keeping the Balance

Sea cucumbers keep the ecosystem in balance. Without sea cucumbers, water becomes **murky** and algae grows more than it should, making it hard for other ocean animals to live. Sea cucumbers prove you don't need a brain to make a difference!

GLOSSARY

algae: plantlike living things that are mostly found in water

animal kingdom: a basic group of natural objects that includes all living and extinct animals

bacteria: tiny creatures that can only be seen with a microscope

ecosystem: all the living things in an area

murky: very dark or foggy

nervous system: the system of nerves in the body that sends messages for controlling movement in the body

skeleton: the bony frame of the body

FOR MORE INFORMATION

BOOKS

Quinlan, Julia J. *Let's Find Out! Marine Life: What Are Sea Invertebrates?* New York, NY: Britannica Educational Publishing, 2017.

Sullivan Rake, Jody. *Faceless, Spineless, and Brainless Ocean Animals: Sea Cucumbers.* North Mankato, MN: Capstone Press, 2017.

WEBSITES

Fact Monster: Animal Kingdom
www.factmonster.com/dk/encyclopedia/science/animal-kingdom
This website has more information about the Animal Kingdom and how scientists classify different creatures.

National Geographic: Sea Cucumbers
www.nationalgeographic.com/animals/invertebrates/group/sea-cucumbers/
This website features more information about sea cucumbers as well as beautiful photographs.

Publisher's note to educators and parents: Our editors have carefully reviewed these websites to ensure that they are suitable for students. Many websites change frequently, however, and we cannot guarantee that a site's future contents will continue to meet our high standards of quality and educational value. Be advised that students should be closely supervised whenever they access the internet.

INDEX